TECH SMARTS

12 QUESTIONS ABOUT ARTIFICIAL INTELLIGENCE

BLACK RABBIT BOOKS

MARNE VENTURA

Table of Contents

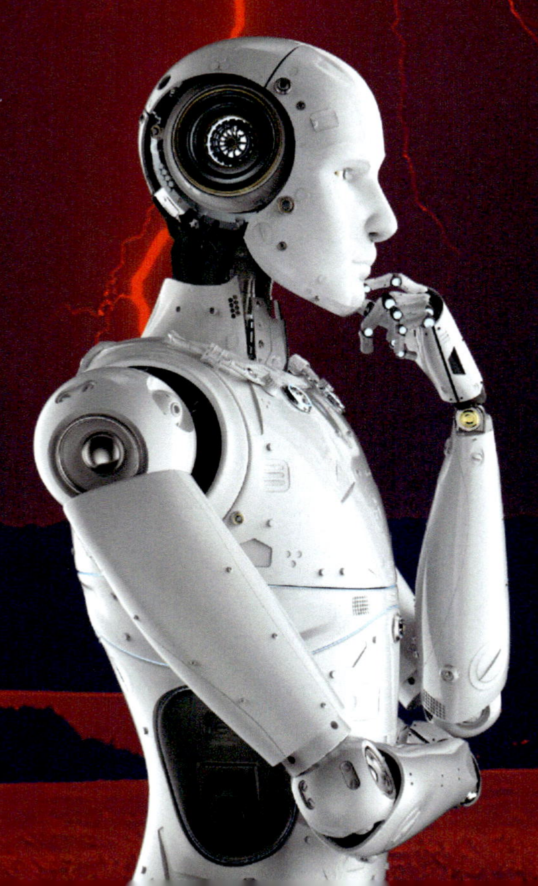

Studying how babies learn could help researchers create better AI tools.

What Is *Artificial Intelligence?*

1

Intelligence is the ability to learn and understand new things. It is the ability to use information to solve problems. It uses reason and logic. Humans are born with intelligence. At six months old, babies learn the sounds that make up words. By age three, they can start to communicate. Babies use the words they learn to solve problems. They can tell a parent what they want. They can share feelings, like anger or love.

Artificial intelligence, or AI, is the ability for a computer to think and learn. AI can do some things like humans. It can understand language. It can solve problems and do tasks. Apple's Siri is one example. A person asks a question to a device. They ask, "What is the capital of Maine?" Siri searches online for the answer. She says, "The capital of Maine is Augusta." She uses a human-like voice.

AI is a powerful tool. Computers can store more data than humans. AI can use this data to solve math problems faster. It can think through a huge number of possible choices. Then it can pick one based on a set of rules. This can be done in seconds.

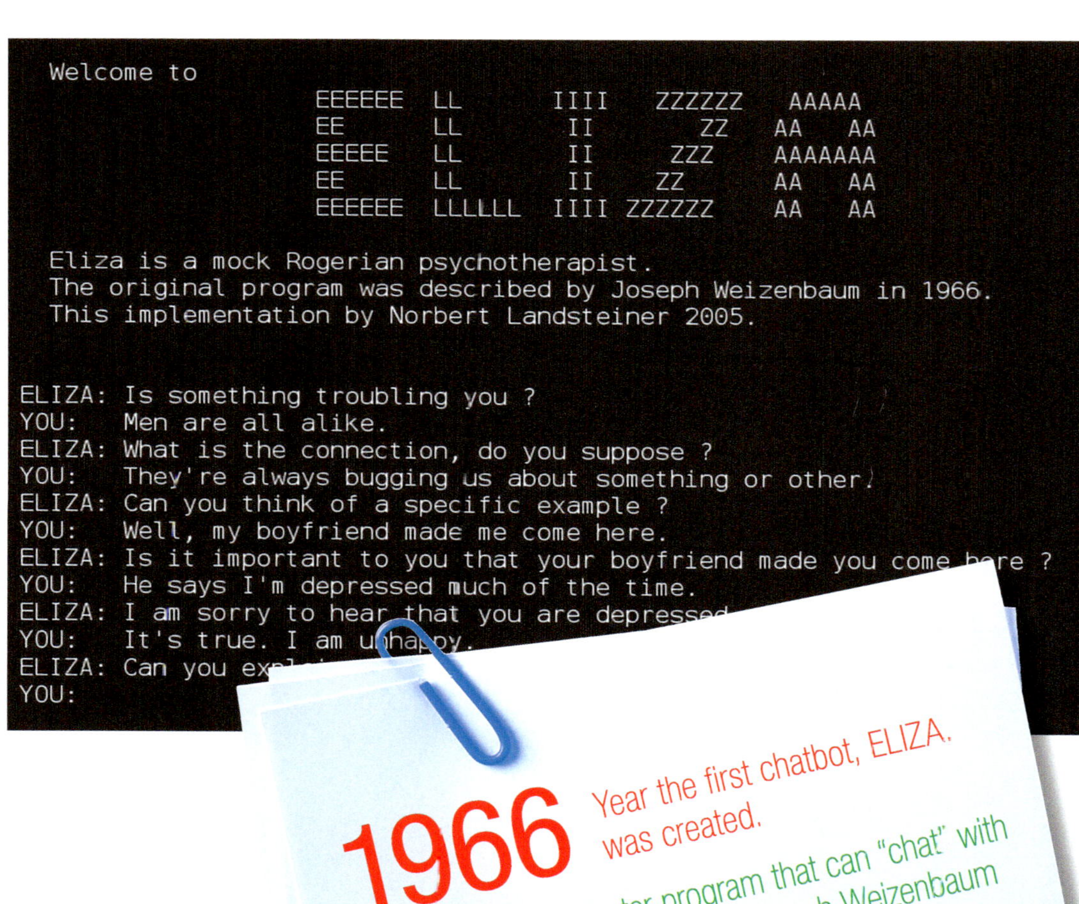

1966 Year the first chatbot, ELIZA, was created.

A chatbot is a computer program that can "chat" with humans. • Computer scientist Joseph Weizenbaum made ELIZA. • A person types words into a computer and ELIZA displays written answers.

Apple's Siri was launched on October 4, 2011.

When Was *AI Created?*

2

Alan Turing was a British mathematician. In 1935, he came up with the idea for a computing machine. It stored information. It followed instructions to do tasks. It even learned things and solved problems. Turing's idea set the stage for electronic computers.

The first electronic computers came out in the 1940s. They followed orders. But they had no memory. They were huge and costly. As time went on, engineers found ways to make computers smaller and faster. They cost less. In 1956, a group of researchers came up with a new computer program. It copied human problem-solving skills. One of the researchers was John McCarthy. He created the term artificial intelligence.

Since that time, AI has grown. Robots do work that people used to do. Some look like people. They can

AI robots can be programmed to play games.

1952 Year the first AI program was run in the United States.

The first program was a checkers game. • Sony made the first robot pet dog, AiBO, in 1999. • GPT-3 was introduced in 2020. It is a tool for AI conversation.

move, see, and talk. AI cars use cameras and **sensors** to move around. In 1997, an IBM computer made history. It won a match against the world chess champion. AI can recognize voices. This **software** led to Apple's Siri and Amazon's Alexa. AI can now make music, paint pictures, and write poetry.

Amazon's Alexa is voice-activated through phones and Echo devices (right).

ROBOTS IN FICTION

Science fiction writers set the stage for AI long ago. Many authors wrote about machines that could think. One was a character in the book *The Wonderful Wizard of Oz*. His name was Tin Woodman. In 1920, Karel Capek used the word robot in a play. In 1968, a movie showed a computer named Hal taking command of a spaceship. The movie was *2001: A Space Odyssey*.

How Does *AI Work?*

3

AI follows a program. Computer scientists write these. AI systems get data from online sources. For example, the AI system for Facebook tracks what a user likes and follows. It also sees what groups they've joined and who their friends are. Based on this data, the program picks what content to show. Each user sees something different.

An AI's program is an **algorithm**. It is set up as step-by-step instructions. For example, an algorithm is used for autocorrect. The AI system compares the typed word with a word in a dictionary database. It finds misspelled words. Then it corrects the spelling. Algorithms are designed to keep learning. As they find more data and do more tasks, they get better. This is machine learning.

Computer vision is a form of machine learning. Computers learn to recognize images. For instance, the AI system in a smartphone takes in the owner's face image. It makes a map of the face. It uses data, like the distance between the eyes and the length of the nose. It stores this data. Then the algorithm compares the map to the owner's face. If they match, the phone unlocks.

3 Types of results created by an AI algorithm: predictions, generation, and analysis.

AI can make a prediction about weather or behavior. • AI can generate, or create, text, images, or a song. • AI can **analyze** trends in data and recommend a solution.

Any camera can be used for face-recognition programs.

How Is AI Used in *Business?*

4

AI systems make it easier for banks to help customers. Instead of a bank teller, people can use an app. They can upload a picture of a check. The app uses an AI program to scan it. An algorithm "reads" the numbers. It finds the amount, account number, and bank number. It also learns to recognize signatures. Then it searches its database for the correct accounts. This happens quickly. And people don't have to wait until the bank opens.

Transportation companies use AI in many ways. Lyft drivers use AI algorithms. They show how long the driver will take to reach a destination. The app finds the best route. It reroutes in case of a traffic jam. FedEx uses an AI-powered sorting robot in China. The robotic arm can sort up to 1,000 packages per hour.

In healthcare, patient records are stored online. Doctors call up this data from a computer. AI systems help doctors

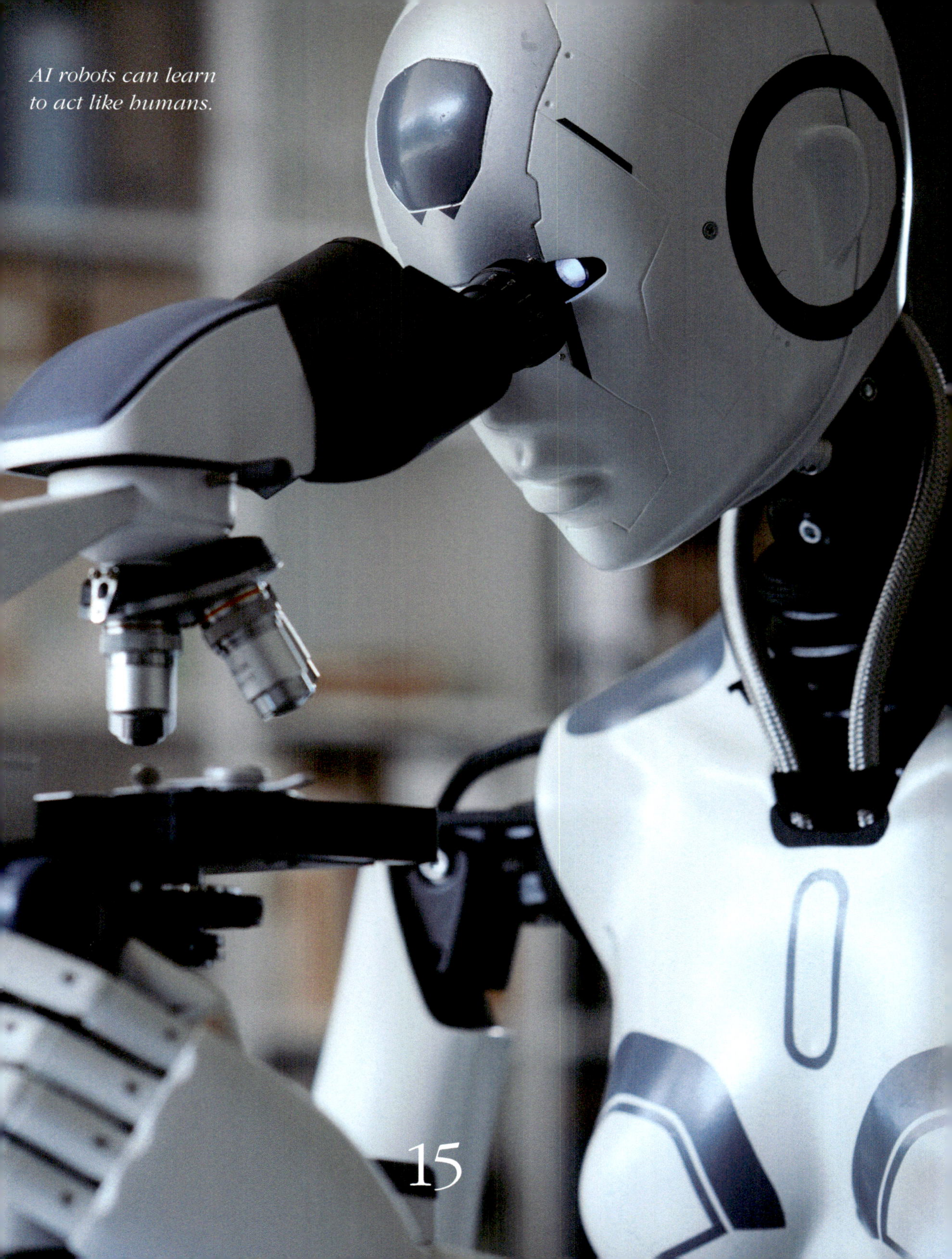

AI robots can learn to act like humans.

find the best way to treat patients. They search data about illnesses. They look for treatments faster than a person could. This helps the doctor make a plan. Devices such as smartphones use sensors to track a patient's heart rate. They alert the patient and doctor if there is a problem.

AI keeps your bank information secure when using contactless payment options.

35 Percentage of companies worldwide that use AI in their business.

AI can advise how long a patient should stay in the hospital. • Banks use AI to find the odds of someone repaying a loan. • AI can count traffic accidents. This data can be used to make intersections safer.

STRENGTH AND POWER Sebastian Thrun is a computer scientist. He works in California. He thinks AI will make humans better. He says, "Just as machines made human muscles a thousand times stronger, machines will make the human brain a thousand times more powerful."

How Is AI Used by *Everyday People?*

5

Chatbots are computer programs that copy human conversation. Chatbots use AI. They use natural language processing (NLP). NLP understands user questions. It finds and gives answers. Siri and Alexa both use NLP. Online shoppers are often helped by chatbots. A customer types a question into a chat box. A bot responds. It may have a human picture. But it's really a computer giving the answers. It is getting harder to know a real person from a chatbot. Experts believe people won't know the difference by 2029.

Amazon Echo device

AI is used in map apps. Drivers, bikers, and walkers use these to find their way. One example is Google Maps. This AI system has a database of maps. Google Maps navigates by using satellites and **aerial** photography. It keeps track of traffic data. It uses cameras to update its data. Travelers ask for directions to a place. The app provides a map with a route. It uses speech to tell a driver when and where to turn. It predicts the user's arrival time.

AI is changing the way people drive cars. Self-driving cars, like people, need to see, hear, and react to what is around them. They need to make fast decisions. Some experts think that most future cars will be driven by AI. The newest cars have some of these features. But fully self-driving cars aren't for sale yet. Experts think they will hit the market in 2035.

A map app finds directions on a smartphone.

Most newer cars come with a built-in screen to see map directions.

300 million Number of lines of code in a self-driving car's program.

Self-driving cars have sensors that find objects around them. • The sensors calculate distances and lane markings. • Video cameras see traffic lights, road signs, and hazards.

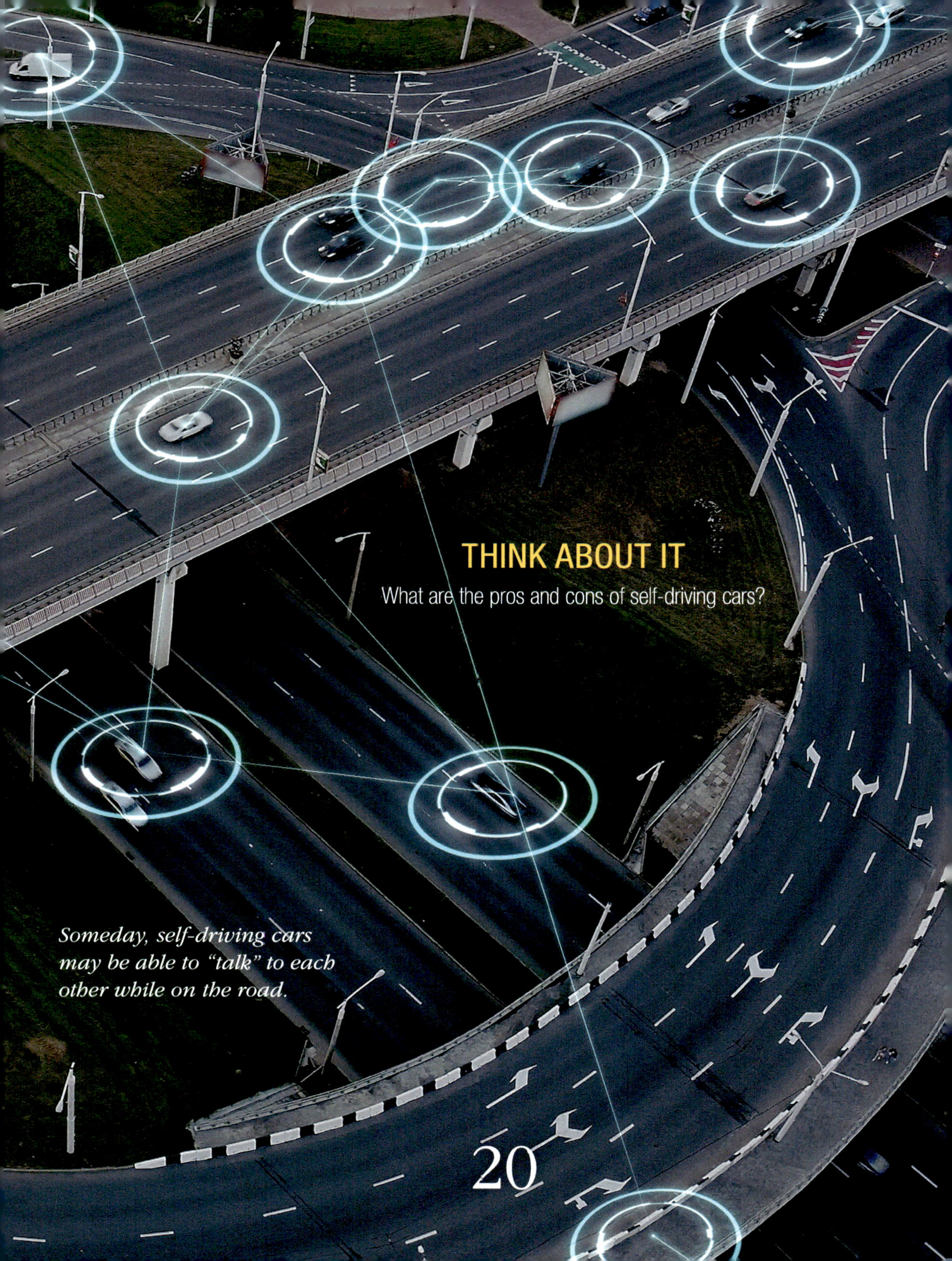

THINK ABOUT IT

What are the pros and cons of self-driving cars?

Someday, self-driving cars may be able to "talk" to each other while on the road.

How Can Students *Use AI?*

6

AI is a powerful tool for learning. Students can find information using a search engine. Learning apps, such as Duolingo, use AI too. Users learn a new language. They take quizzes. The app adapts the lesson to meet the student's needs. Students can improve their writing and math skills with learning apps. They can learn about science and technology. They can study history. They can learn about arts and culture.

At the end of 2022, **generative** AI came out. It makes new content as if it were human. ChatGPT is an example. It can write an essay on any topic. It can write a poem or story. Other AI systems create pictures or write music in response to a prompt.

AI can function as a tutor. Human tutors help students learn or better understand subjects. AI tutors tap into a vast online database. They can answer questions and explain concepts quickly. Blueprint Test Prep is one example. It helps college students prepare for exams. Age of Learning is another tool. It helps students from preschool to second grade master reading and math skills.

AI can adapt lessons to fit a student's individual needs.

PLAGIARISM Plagiarism is using someone else's words or ideas as if they were your own. It is illegal. But it is easy to do in today's online world. When using online sources, you need to give credit. You should cite the original creator. This goes for quotes, music, and even dance moves. You should also get permission to use another person's content.

30 Percentage of fifth-grade students whose math grades improved after using an AI tutor.

The AI system helped the kids learn multiplication and division. • The students were 20 percent less anxious about math. • The system increased their motivation to learn.

7 How Is AI Helpful to *People?*

AI makes life better and easier. It is especially helpful for people with disabilities. People in wheelchairs use apps to find places with ramps and elevators. People with physical **impairments** control computers with speech. People with impaired sight have apps that help them see. The apps read text. It describes the surroundings and objects. AI is used in all these things.

AI is great at doing simple tasks over and over again. This means people don't have to do them. It frees up their time to do more meaningful work. Some teachers grade papers and tests with AI. AI also does tasks that are unsafe for people. Warehouses use robots to lift and move boxes. This prevents human workers from getting hurt. Plus a robot can lift heavier boxes. It can do bigger jobs faster.

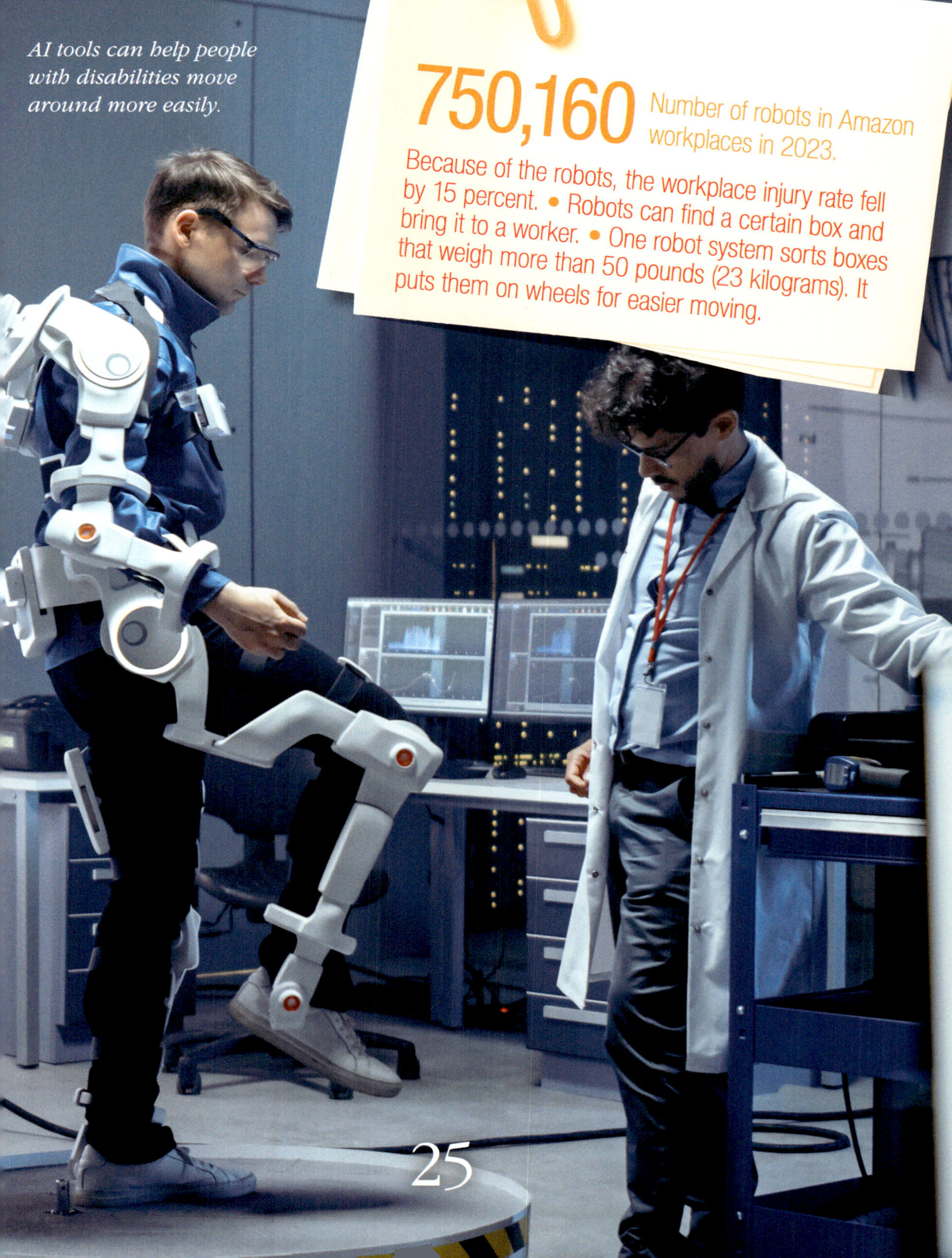

AI tools can help people with disabilities move around more easily.

750,160 Number of robots in Amazon workplaces in 2023.

Because of the robots, the workplace injury rate fell by 15 percent. • Robots can find a certain box and bring it to a worker. • One robot system sorts boxes that weigh more than 50 pounds (23 kilograms). It puts them on wheels for easier moving.

AI is helpful for researchers. Huge amounts of data are stored online. This is called big data. AI sorts through this faster than a person can. It helps researchers find and use information more quickly. Business owners use it to decide which products will sell the best. AI predicts where wildfires might break out. This helps fire fighters and residents prepare. They catch the fire before it gets too big. AI can save lives and prevent damage.

Prosthetic limbs can be controlled with AI.

How Is AI Harmful to *People?*

8

As AI becomes more powerful, people worry it could pose a risk. Jobs done in the past by people are now done by robots. New jobs in technology are growing. But people in factories may not have the knowledge for these new jobs. They would be out of a job. Some experts worry the job market will be hurt by AI. People replaced by robots might not find other jobs.

Fake news and media bias are dangers of AI. People need true information about the weather and news. This helps them stay safe. It helps them make decisions, like how to vote. Some people use AI to change pictures, video, and audio. They post false information. This can be harmful in many ways. An innocent person can be made to look guilty. The way a person votes could be swayed by false data. People could be scared about an event that is not real.

There is danger in AI weapons as well. Drones help the military. They can go into places that are dangerous. They get information or find mines. This keeps a service person from getting hurt. A person controls these types of drones. But new AI weapons don't need human control. They find and destroy targets on their own. Experts worry about this. Powerful AI weapons could kill innocent people. A **hacker** might take control of AI weapons.

700 Number of risks to society from AI, identified by scientists at Massachusetts Institute of Technology (MIT).

GPT AI systems compete with humans. • Experts worry these systems might become more powerful than people. • Harmless AI tools could be reprogrammed to do harm.

A factory uses AI robots to build cars.

THINK ABOUT IT

Think about the jobs AI will replace. Who will be affected by these changes the most? How might this affect wages and the job market?

What Are Safety Guidelines *for Using AI?*

9

When people use AI, it leaves a digital footprint. This is a record that AI systems keep. It tracks online searches and purchases. It tracks downloads and views. Some apps know the location of the user. These are all things that strangers should not know. To protect your privacy, turn off location services. This is a setting on most devices. Make sure search engines are secure. Block ads. Avoid websites that ask you to accept all cookies.

Be **skeptical** about information you read online. The internet is a valuable tool. You can do research, get news, or connect with friends. But AI makes it easy to post fake pictures, videos, or audio files. Learn to identify **reliable** sources of information. Have you heard of the source before? Can you find the same facts on more than one source? Does the website list sources for the information?

Companies are using AI to track more of your online activity.

AI learning tools can be great resources. But students should double check that all facts are true. Then, write the final work in your own words. It is tempting to use AI, but it is not **ethical**. It is a form of cheating. Some schools have rules against AI use. You could get in trouble if you use it on assignments.

THE RIGHT PROMPT

Some teachers are embracing ChatGPT as an important learning tool. They are teaching their students how to write better prompts. They help them use the right words. They help them check facts. They show them how to think **critically** about the answer given.

AI can summarize articles for a quick overview of a topic.

51 Percentage of young people that have used generative AI.

The survey included people ages 14 to 22. • The most common use was to get information and brainstorm ideas. • Only 25 percent never used AI for schoolwork.

What Are Some Pros and Cons *of AI?*

10

AI systems reduce human error. For example, AI weather forecasting is more accurate than human forecasters. AI-based systems can analyze medical data. They help doctors diagnose diseases. AI **simulations** can train new workers. This helps prevent errors. It gives workers a chance to correct mistakes safely.

AI can save businesses money. When robots do repetitive work, employers save the cost of hiring people. AI can analyze a company's cash flow. It can figure out how to spend less and earn more. But there is a downside. Cutting human labor can cause a negative effect on people's lives. The workers might not have the skills for other work. This could lead to more joblessness and poverty.

Many reports show that AI will create new jobs. Companies will need to hire new roles. They will

AI use in the medical field may help save lives.

need robotic engineers or data scientists. More AI researchers will be needed too. These workers will keep the AI tools running. But they won't replace the old jobs. These new jobs will require different skills and education. Some workers can get new training to do this kind of work. Others may not have the skills to compete in this new job space. They will simply be replaced.

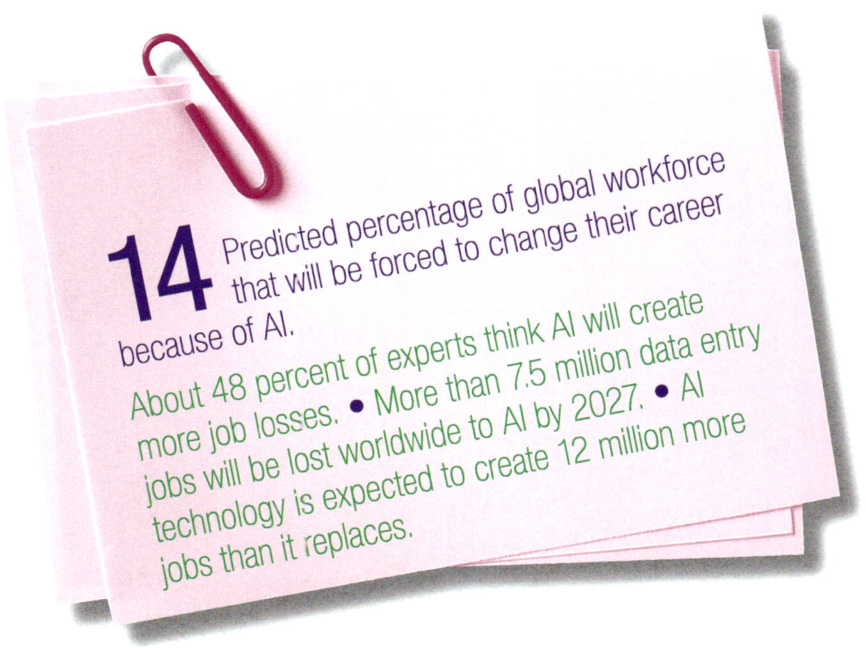

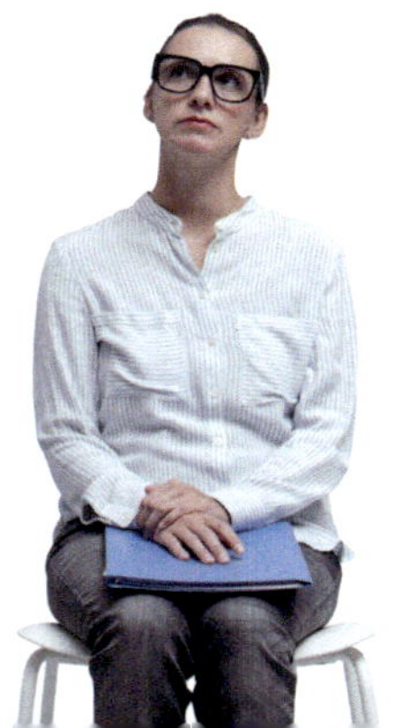

Workers may compete with AI tools for jobs.

Are There Laws about *AI Use?*

11

AI has become very powerful in recent years. Laws in the United States have not kept pace with AI. Experts and US leaders agree that more laws are needed to govern the use of AI. Data privacy is a concern. AI systems collect data from users. People have a right to keep things like their bank or medical information private. This information could be used unlawfully. Some state laws protect people's data. But there is no national law on this yet.

The Children's Online Privacy Protection Act (COPPA) was passed in the late 1990s. It says that personal information about kids under age 13 cannot be collected or used. This applies to websites for children. Most social media sites have an age requirement. Kids need to be at least 13 years old.

In 2023, President Joe Biden made an order. He asked for new AI standards. He wanted better safety and security. He called for AI testing. The systems should not pose threats to public safety. He wanted it clear when AI tools were being used. He also asked for safety checks on AI military weapons.

Military robots are quickly moving from the design stage to the battlefield.

45 Number of US states that introduced bills to pass AI laws in 2024.

Colorado passed laws to stop discrimination. • Florida gave schools grant money to use AI to help students and teachers. • Indiana has an AI task force.

Apps and programs can protect from AI data breaches.

What Is the Future *of AI?*

12

Experts cannot fully predict what the future will hold for AI. One expert is Terrence Sejnowski. He is a neurobiology professor at the University of California in San Diego. He says AI is evolving fast. It is quickly changing how we use technology. He compares it to when the Wright brothers first made an airplane with a motor in 1903. People could not imagine all the ways that flight would change the world. That is like AI today.

Wright brothers on a US stamp

New AI tools work like the human brain. They can learn. They may soon be able to set goals. Some could have long-term memory. This makes them unpredictable.

Future AI could bring harm. People might become too dependent on technology. Robots might lead to less human contact. There are environmental concerns. AI computer data centers use large amounts of electricity. The centers could lead to greater carbon **emissions**. These cause global warming.

AI also holds the power to make life better. Experts think people will use more AI in their daily lives in the next decade. AI systems will be personal assistants and tutors. They could be career counselors and therapists. There may be AI accountants or lawyers. AI will write computer code and build and sell products. It will help customers and make decisions. AI is an amazing tool. It has the power for both good and bad outcomes. Like any tool, its worth depends on how people use it.

NVIDIA is one of the most advanced platforms for generative AI.

Wilbur and Orville Wright changed the world the day they flew the first airplane.

THINK ABOUT IT

Look up AI predictions from years past. Were they correct? Compare them with new predictions today.

Future airplanes may be flown by an AI pilot.

Safety Tips

Understand Privacy Settings

Take time to review settings on all your accounts. Be sure you understand what each setting means. If not, ask for help. A trusted adult can go through it with you. This is especially important for privacy. Opt out of sharing your data.

Think Critically

Be careful about chat messages from unknown accounts. They could be from a bot. Report any suspicious content. Don't believe all images and posts are real right away. Think about who is posting the information. Is it a good source? Are the results biased? Do your own research to make sure you have all the facts.

for AI

Don't Rely on AI

AI is a great tool for learning. But it shouldn't do all the thinking for you. Don't rely on AI to make decisions. Those should be made by you after careful thinking. AI does not have emotional understanding. It cannot replace human relationships.

Stay Informed

Learn about the latest AI tools. Understand their uses and possible risks. AI technology changes quickly. By being up to date, you can make smart choices. You can use AI responsibly.

Glossary

aerial
Performed in the air.

algorithm
A set of steps that are followed in order to complete a computer process.

analyze
To study something closely and carefully.

critically
In a way that involves careful study of its merits and faults.

emission
Something that is released into the air.

ethical
Following accepted rules of behavior.

generative
Relating to or capable of creating something new.

hacker
A person who illegally gains access to and sometimes makes changes within a computer system.

impairment
A condition in which a part of a person's body or mind is damaged and does not work well.

reliable
Able to be trusted or believed.

sensor
A device that detects light, movement, or other information.

simulation
Something that is made to look, feel, or behave like something real.

skeptical
Having or expressing doubt about something.

software
The programs that run on a computer and perform certain functions.

For More Information

Books

Allen, John. *Exploring Careers in AI.* San Diego: ReferencePoint Press, Inc., 2025

Harris, Chris G. *Understanding Generative AI.* Buffalo, NY: PowerKids Press, 2025.

Idzikowski, Lisa. *The Challenges of AI.* Minneapolis: Lerner Publications, 2025.

Kuehl, Ashley. *Artificial Intelligence.* Minneapolis: Bearport Publishing Company, 2025.

Websites

A Kid's Guide to Artificial Intelligence and Machine Learning
www.aiprm.com/education/a-kids-guide-to-artificial-intelligence-and-machine-learning

Will Future Robots and AI Take Over?
education.nationalgeographic.org/resource/will-future-robots-and-i-take-over

The Future of Artificial Intelligence
thecrashcourse.com/courses/the-future-of-artificial-intelligence-crash-course-ai-20

About the Author

Marne Ventura is a children's book author and a former elementary school teacher. She holds a master's degree in education with an emphasis in reading and language development from the University of California.

Index

TOP RANK is published by Black Rabbit Books, P.O. Box 227, Mankato, MN, 56002.

• Designed by Danny Nanos • Photographs © Dreamstime/Robhainer, 4, Sasinparaksa, 39, Shao-chun Wang, 10; Getty Images/dowell, 13, SolStock, 23, xia yuan, 28–29, 38, Yuichiro Chino, 14; Internet Archive/NASA, 42; Shutterstock/AlinStock, 20, Alliance Images, 27, Andrey Suslov, cover, 1, Beautrium, 35, Belinda Pretorius, 48, BLACKDAY, 26, Bobkov Evgeniy, 18–19, Charles Brutlag, 17, DedMityay, 7, 44, dennizn, 33, Gorodenkoff, 25, Ground Picture, 21, 22, Gulpa, 10, Gumbariya, 11, Hepha1st0s, 41, Jakob Berg, 9, Jolygon, 46–47, Kongpraphat, 12, Lamai Prasitsuwan, 12, New Africa, 16, NicoElNino, 37, 44, Nikolay Zaborskikh, 2–3, one photo, 18, Phonlamai Photo, 2, Rawpixel.com, 31, Scharfsinn, 24, selinofoto, 45, Skycolors, 43, spatuletail, 8, Stock-Asso, 15, Stokkete, 36, Teerapong mahawan, 5, Tony Baggett, 40, Yeti studio, 32, 45; Wikimedia Commons/public domain, 6

Library of Congress Cataloging-in-Publication Data: Names: Ventura, Marne, author. | Title: 12 questions about artificial intelligence / by Marne Ventura. | Other titles: Twelve questions about artificial intelligence | Tech smarts. | Description: Mankato , MN: Black Rabbit Books, [2026] | Series: Tech smarts | Includes bibliographical references and index. | Ages 9–13 | Grades 4–6 | Identifiers: LCCN 2024054648 (print) | LCCN 2024054649 (ebook) | ISBN 9781644668146 (library binding) | ISBN 9781644668467 (paperback) | ISBN 9781644668788 (ebook) | Subjects: LCSH: Artificial intelligence—Juvenile literature. | Classification: LCC Q335.4 .V46 2026 (print) | LCC Q335.4 (ebook) | DDC 006.3—dc23/eng/20241127 | LC record available at https://lccn.loc.gov/2024054648